# Land and Water
## in Hawaii

by Susan Halko

# Contents

# Science Vocabulary

**natural resource**
A **natural resource** is a part of Earth that people use.

Water is a **natural resource.** People fish in water.

**ocean**
The **ocean** is a large, deep body of salt water.

The islands of Hawaii have water all around them. This water is the **ocean.**

## mineral
A **mineral** is a nonliving material found in nature.

A **mineral** made these rocks black.

## soil
**Soil** is the top layer of Earth's land made of rocks, minerals, and dead plants and animals.

Farmers in Hawaii grow crops in **soil.**

## volcano

A **volcano** is an opening on Earth from which lava flows.

**Volcanoes** can cause fast changes on Earth.

## earthquake

An **earthquake** is a sudden shaking of the ground caused by land moving.

An **earthquake** caused a lot of damage to this road in Hawaii.

## weathering

**Weathering** is the breaking apart of rocks.

Moving water can make rocks smooth. This is a kind of **weathering.**

## erosion

**Erosion** is the movement of rocks or soil caused by wind, water, or ice.

Water and wind can move rocks and soil. This is a kind of **erosion.**

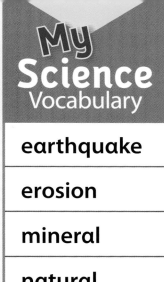

My Science Vocabulary

earthquake

erosion

mineral

natural resource

ocean

soil

volcano

weathering

# Hawaii

Hawaii is a group of islands. The islands are in the **ocean.** Each island in Hawaii is different.

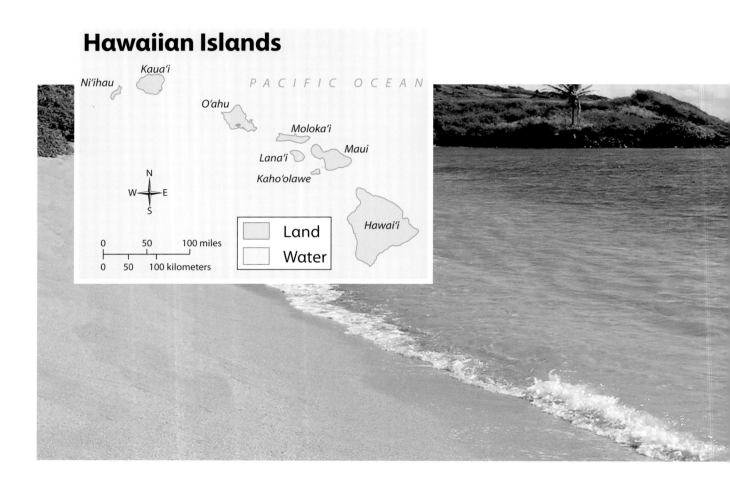

**Hawaiian Islands**

Ni'ihau · Kaua'i · O'ahu · Moloka'i · Lana'i · Maui · Kaho'olawe · Hawai'i

PACIFIC OCEAN

N W E S

0   50   100 miles
0   50   100 kilometers

Land
Water

**ocean**

The **ocean** is a large, deep body of salt water.

There are many types of land and water on these islands. There are sandy beaches and grassy fields. There are large lakes and small streams.

Some islands in Hawaii have tall mountains.
The mountains soar into the sky.

These mountains are on the Na Pali Coast in Kaua'i.

Some islands in Hawaii have rain forests.
Plants and trees grow in the rain forests.
Fish swim through the rivers and streams.

Trees grow in a rain forest in Maui.

# People Use Water

Hawaii has many **natural resources.**
Water is a natural resource. People use
ocean water to fish, swim, and surf.

**natural resource**

A **natural resource** is a part of
Earth that people use.

People drink fresh water. Some plants grow in fresh water. Fresh water is not salty.

Taro plants need a lot of fresh water to grow.

# People Use Land

**Soil** is a natural resource. Farmers in Hawaii use soil to grow plants. People eat parts of these plants for food.

Farmers grow sweet pineapples in soil.

**soil**

**Soil** is the top layer of Earth's land made of rocks, minerals, and dead plants and animals.

People in Hawaii use plants in other ways, too. They use bamboo and flowers to make things.

— bamboo fishing pole

necklace — of flowers

Rocks and **minerals** are natural resources.
Long ago, people in Hawaii built rock walls.
Today, people cook food over hot rocks.

**mineral**

A **mineral** is a nonliving material found in nature.

The rocks on this beach are basalt.

Minerals give basalt its black color.

# The Land Changes

The land in Hawaii can change quickly
or slowly.

A **volcano** can cause fast or slow changes.
Volcanoes formed the Hawaiian Islands.
This took millions of years!

**volcano**

A **volcano** is an opening on Earth
from which lava flows.

Each Hawaiian island formed in the same way. A volcano erupted on the ocean floor. The lava cooled and turned into rock. It formed a mountain.

lava

A volcano erupted in the ocean near Hawaii.

The volcano kept erupting and growing. Soon the mountain reached the top of the water. It became an island.

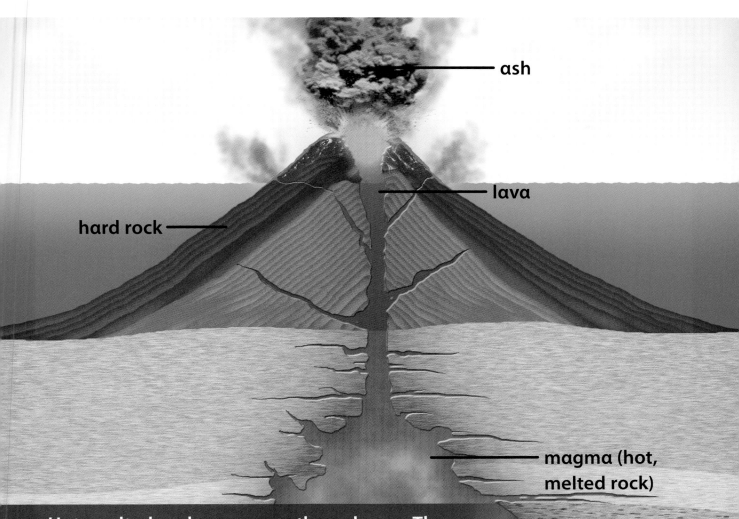

ash

lava

hard rock

magma (hot, melted rock)

Hot, melted rock comes up the volcano. Then lava flows down the sides of the volcano.

Volcanoes can also cause fast changes. Volcanoes explode ash into the air. This can block the sunlight.

The ash is good for soil. It can help plants grow.

Plants grow in soil and ash from a volcano.

An **earthquake** can also cause fast changes.

An earthquake broke apart this road.

**earthquake**

An **earthquake** is a sudden shaking of the ground caused by land moving.

Earthquakes can shake the ocean floor. An earthquake under the ocean can cause huge waves to form.

**Waves from an earthquake flooded this land in Hawaii.**

Water and wind cause Hawaii's land to change slowly. Water from this waterfall rushes over rocks. Over time, it makes the rocks smooth. This is a kind of **weathering.**

**weathering**

**Weathering** is the breaking apart of rocks.

Water and wind move rocks and soil.

This is a kind of **erosion.**

The land is always changing in Hawaii!

**erosion**

**Erosion** is the movement of rocks and soil caused by wind, water, or ice.

# Conclusion

Hawaii is a group of islands. They formed from volcanoes. Hawaii has many natural resources. It has water, rocks, and soil.

The land changes. Some changes happen quickly. Some changes happen slowly.

## Think About the Big Ideas

1. How do people in Hawaii use land?
2. How do people in Hawaii use water?
3. How does the land in Hawaii change?

# Share and Compare

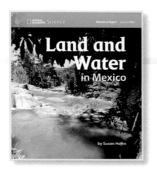  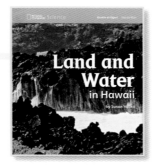

## Turn and Talk

Compare ways people use fresh water and salt water. How are they the same? How are they different?

## Read

Read your favorite page to a classmate.

## Write

Write about your favorite photo in this book. Then share your writing with a classmate.

## Draw

Draw three ways you use natural resources. Show your drawing to a classmate.

## Meet Katey Walter Anthony

Katey Walter Anthony is a scientist. She learns by observing. Katey observes land and water in the Arctic. The Arctic is a cold place. It has ice and snow.

The ice is melting. The land is warming, too. A gas goes into the air when this happens. Katey wants to learn how to use this gas.

# Index

**Acknowledgments**
Grateful acknowledgment is given to the authors, artists, photographers, museums, publishers, and agents for permission to reprint copyrighted material. Every effort has been made to secure the appropriate permission. If any omissions have been made or if corrections are required, please contact the Publisher.

**Photographic Credits**
Cover (bg) Photodisc/Getty Images; Cvr Flap (t), 5 (t), 16 (l) Jupiterimages/Thinkstock/Alamy Images; Cvr Flap (c), 15 (l) Stephanie Maze/Corbis; Cvr Flap (b), 7 (b), 27 Diane Cook and Len Jenshel/National Geographic Image Collection; Title (bg) Steve Raymer/National Geographic Image Collection; 2–3 Corel; 4 (t), 12 (l) Pacific Stock/SuperStock; 4 (b), 8–9 Mike Brake/Shutterstock; 5 (b), 14 PhotoDisc/Getty Images; 6 (t), 18–19 DigitIStock/Corbis; 6 (b) 24 Craig Lovell / Eagle Visions Photography/Alamy Images; 7 (t), 26 Stacy Gold/National Geographic Image Collection; 10 David Alan Harvey/National Geographic Image Collection; 11 Kriss Russell/iStockphoto; 12 (r) Robert B. Goodman/National Geographic Image Collection; 13 Nik Wheeler/Alamy Images; 15 (r) Tomas Del Amo/Alamy Images; 16 (r) Michael S. Yamashita/Corbis; 17 Michael J Thompson/Shutterstock; 20, 28 DigitalStock/Corbis; 22 Steve and Donna O'Meara/National Geographic Image Collection; 23 Bernd Mellmann/Alamy Images; 25 Henry Helbush/National Geophysical Data Center (NOAA); 30, 31 Institute of Northern Engineering, University of Alaska, Fairbanks; Inside Back Cover (bg) Walter Meayers Edwards/National Geographic Image Collection.

**Illustrator Credits**
8 Mapping Specialists; 21 Paul Mirocha

Neither the Publisher nor the authors shall be liable for any damage that may be caused or sustained or result from conducting any of the activities in this publication without specifically following instructions, undertaking the activities without proper supervision, or failing to comply with the cautions contained herein.

**Program Authors**
Kathy Cabe Trundle, Ph.D., Associate Professor of Early Childhood Science Education, The Ohio State University, Columbus, Ohio; Randy Bell, Ph.D., Associate Professor of Science Education, University of Virginia, Charlottesville, Virginia; Malcolm B. Butler, Associate Professor of Science Education, University of South Florida, St. Petersburg, Florida; Nell K. Duke, Ed.D., Co-Director of the Literacy Achievement Research Center and Professor of Teacher Education and Educational Psychology, Michigan State University, East Lansing, Michigan; Judith Sweeney Lederman, Ph.D., Director of Teacher Education and Associate Professor of Science Education, Department of Mathematics and Science Education, Illinois Institute of Technology, Chicago, Illinois; David W. Moore, Ph.D., Professor of Education, College of Teacher Education and Leadership, Arizona State University, Tempe, Arizona

**The National Geographic Society**
John M. Fahey, Jr., President & Chief Executive Officer
Gilbert M. Grosvenor, Chairman of the Board

Copyright © 2011 The Hampton-Brown Company, Inc., a wholly owned subsidiary of the National Geographic Society, publishing under the imprints National Geographic School Publishing and Hampton-Brown.

National Geographic School Publishing
Hampton-Brown
www.NGSP.com

Printed in the USA.
Quad Graphics, Leominster, MA

ISBN: 978-0-7362-5520-2

18

10 9 8 7 6 5